Our Cosmic Evolution

Written and illustrated by

Roy S. Timmreck

Preface

Everything is always changing, so, everything that can do so must continuously adapt and compete, and *evolve* to survive.

When verifiable scientific knowledge is applied to an expansive view of evolution, the understanding that results is enormous. Its implications are very different from what most people are thinking about themselves and their world today.

The refreshing natural vision, conclusions, and implications of *Our Cosmic Evolution* are surprising, dramatic, and will certainly be controversial, but only because we have been immersed for millennia in unscientific beliefs and opinions.

Hello

A vision of science and evolution, this is a book about the splendor of humanity, and it is about you, how important and how amazing you are, you personally, your everyday life, your whole life. Of course, the earth and the universe are really big and you are only one person, but that is only how it seems. Every ordinary good person living an ordinary good life, being kind and strong and creative and helpful is always greater and far more important to the world and the future than all the most famous and powerful unkind people who are or have ever been. When you understand your true nature and the true nature of our earth and our universe, then you will understand; to see that, this is a good place to start.

Evolution

Collisions of light-like stuff during and after the Big Bang 13.8 billion (thousand million) years ago create hydrogen atoms whose combined gravity crushes them together to ignite as the first stars. Those stars make larger atoms and finally supernova, making even larger atoms. Then, gravity and atoms from generations of stars make planets and moons, some suitable for molecules to evolve to life, and for life to evolve into all the microbes, plants, and animals that are and have ever been.

Ever since the earth cooled and its outer crust formed 4.5 billion years ago, gravity and the flow of molten rock below the surface have caused the continents to collide and separate and slide around. So the continents change, the oceans change, and life in the water adapts to a changing variety of environments. Meanwhile, those living things that evolved to capture energy from sunlight are consumed by those who didn't, as adaptation, predator, and prey lead to an unimaginably rich explosion of life forms and further evolution.

Blue green algae produce oxygen gas changing the atmosphere of the planet, protecting it from cosmic rays. Tiny plants populate the land; insects populate the air. Fish crawl out of the water, but their fins are poorly designed for walking. Reptiles evolve legs, better than fins at walking, but no match for the evolution of dinosaurs that can walk and run beautifully. Dinosaurs however, have a serious vulnerability: mammals can outmaneuver them by running in packs, distracting them, and destroying their eggs.

So here we are on our spinning earth, orbiting our sun at 18.6 miles a second, which pulls us with it around the center of our Milky Way galaxy at 140 miles a second, which in turn is spiraling through space in the direction of the constellations Leo and Virgo at 360 miles a second. You have probably gone over 1500 miles while reading this sentence.

Each and every one of us is constructed of about 7,000,000,000,000,000,000,000,000,000 atoms, 62.9% of which are hydrogen, older than the stars, from the Big Bang itself. The rest were made in a family of stars that burned for billions of years exploded and threw them into space before our sun was born, atoms that have already been recycled through rocks, mud, air, and water, as well as many other people, plants, and animals alive and dead.

Each of us began as a single cell 0.12 millimeter in diameter, only able to survive in seawater. As we grow to 30 trillion cells, our height increases to an average of 5 feet 7 inches (1.7 meters), an increase of 14,167 times. Our cells contain mitochondria that process most of our energy and probably originally evolved as separate organisms. The average lifetime of our cells is 7 years, some muscle cells live 15 years; we shed over 55,000 skin cells every *second*. Trillions of microorganisms including viruses, bacteria and fungi live on and in us, many helping us process our food. From every inhaled breath, we take sextillions of oxygen molecules that used to be trees, shrubs, flowers, weeds, and grass around us near and far, and then exhale, fewer, but still sextillions of carbon dioxide molecules that become that same whole ecology of plants around us near and far.

Hail, Hail, the Gang's All Here

We have evolved, our form has changed and adapted, our brains have gotten bigger, our knowledge has adapted, competed, and evolved, but we do not suddenly pop into existence fully evolved, free of all our previous evolution. A modern axe does not have a stone axe at its core; automobiles contain horsepower, but no actual horses. We on the other hand, do actually physically and mentally still possess much of our entire evolution. Starting as a single cell in the special sea of our mother's womb, we replay how we evolved over billions of years in the ancient sea. Before birth we have been one cell, grown to many cells, been unable to see and developed eyes, been unable to hear and developed hearing, and also developed arms and legs to prepare for our lives on land. Only at birth do we breathe air. We learn to roll over, to crawl, to walk, to talk better and better, to think better and better.

We see in children the evolution of memories and feelings that gave us our first levels of thinking, of making sense of our world. As they grow, feelings and believing evolve to include thinking logically and seeking ways of knowing what is true. But that road is bumpy; the less evolved levels of our mind are still there, older, more established, already tested by adaptation and survival. Our most recently evolved mental capacities are smarter, but not yet as deeply ingrained in our physiology. We have only been human for a very brief fraction of our evolution.

The fact that in so many ways, our whole evolution is still with us challenges our current social evolution. *Feels good, feels bad,* and *eat* have been evolving for *billions* of years, from the very beginning of life on earth. It is not surprising that we can still eat as if it is our single purpose. We can also still think and act as amorally as our fish ancestors or with the heartless ferocity of our reptile ancestors. Our newer most highly evolved thoughts and behaviors lie upon, coexist with, and are challenged by their own older deeper still powerful survival-established versions. Consequently, our instincts, drives, emotions, and beliefs may override our higher values, clear thoughts, better judgment, and better logic, not because of any urging by evil spirits, but because they are so old within us, because they were our ways, some for billions, others for hundreds or tens of millions of years.

Our old instincts and drives come from times in our development and long ago environments where they served us well. But modern society is a very different environment from dinosaurs at the entrance to our burrow, or eking out a living in the wild through the winter, or competing with dangerous animals for territory. We find ourselves still wanting to be the important one, to make others do what we think should be done, we still want the toy the other child has, we still have impulses to take by force what we want. We are challenged to manage those old tendencies in ways that are positive, beneficial, and up to our current level of evolution, but depending on our genes, childhood, and life experiences, that can be a big problem.

Battle of the Berry Patch

As our world shrinks, the original problem of the berry patch just gets bigger and bigger. Our village relies on that berry patch every year when the berries are ripe. Then one year, the berries are gone; people from a different village have come and picked them all. After years of strife, our villages join together for whatever reason, solve that berry problem, and go on to grapple with other things. As long as nations struggle over resources, partly for the needs of their people, but more for the ego, power, greed, beliefs, status, and perceived glory of their elite, the berry patch problem continues.

How may we ever solve inequities in freedom, status, education, opportunity, and resources?

A great part of that problem is the evolution of leadership. At first, the leader was whoever was biggest and strongest, but quickly, intelligence, cunning, knowledge, experience and wisdom became more important. For millennia, societies had small upper classes whose superior status, education, wealth, connections, and resources properly gave them a monopoly on who was best qualified to lead. As farmers have gone to the cities, as economies and cultures become world-wide, as communication skyrockets, as human knowledge has grown exponentially, and as many millions become educated, that old monopoly is no longer valid. Also, *now* the challenges of government are so large and complex that competence really requires an intelligently designed structure of truly dedicated professionals highly educated for the jobs they are to do.

Clearly we are not nearly there yet. We still have warlords leading nations, charismatic leaders and leaders who come from money and status, but who lack experience and maturity, leaders supported by extremists, and even leaders who are excellent in some spheres, but aren't up to the full breadth of the challenges they face.

It is no coincidence that all the current forms of government, all so sure they are the best, tend to have extremely wealthy leaders. Someday, leadership competence will just be about doing best by mankind and the environment. Now, however, it still includes the need for might and wisdom in the continuing battle of the berry patch.

What is needed is a single world government comprised of people of integrity who truly care about and have worked and studied to competently strive to best serve all the people of the world. Even then, the problem must be solved that only a small percentage of those most truly dedicated in any challenging profession are impressively good at what they do.

Many leaders, social systems, and religions have killed thousands and some have killed millions of good people to impose their beliefs and visions of how things should be—and they're still doing it! It would seem that we are hopelessly far from being as evolved as we should be. Yet there is hope in the unassailable truth that for thousands of years now, a great many people in all societies, primitive and modern, at all levels of education, have been and are truly good people who regard everyone as their great family and the ecology of the earth as their charge to manage, respect, and protect.

I believe, therefore, I am.

Though many smart animals can know and remember many things, communicate with their own as well as they need to, and have a variety of emotions, their thoughts do not include explanations of what they know, feel, imagine, or do not understand. As we humans evolved past that level, we have struggled with the challenge of explaining why things do or do not happen. We learned to recognize correlation, this happened, and then that happened, but still presume it is cause and effect more often than it is.

As we developed more discerning thinking and logic, we got better at determining truth, but our tendency to imagine and believe spiritual explanations for things that we don't understand has been a massive problem. Our history shows that societies that lack scientific knowledge of the physics, chemistry, and biology of us, our earth, and our universe can believe a great variety of things that are false, and will often feel it necessary to oppress or even kill those who disagree with them.

Predictably, their brutality has its origins in survival behaviors. Single families, as the oldest social groups, particularly vulnerable to attack, develop a healthy fear of and readiness to kill outsiders who want to take their possessions. Larger groups have the same problem on a larger scale. The evolution of whose life matters and whose doesn't expands in general as societies get larger and become more diverse. All along the way, the lives of the in-group are respected; the lives of the out-group matter less, and often don't matter at all.

Now, the in-group can be our family, our village, tribe, or nation, our religion, our military, our political views, our form of government, our race, our culture, our heritage, our clique, and most important of all, those who agree that *we are right*. The problem is very large and complex because those groupings mindlessly disregard more important criteria such as who are the good people.

Frightened, paranoid, authoritarian governments, leaders, and their followers, throwbacks from a time of dictating truth rather than grasping it, are quick to suppress and use deadly force on an out-group that may be defined as: *You aren't supporting my lies* or even, *You aren't supporting my lies with sufficient enthusiasm*. They are paranoid because they realize that their enemy is truth itself, which includes all those who know the truth as well as all those who may find it out or figure it out. Invariably of course, their closest insiders know the truth, so they are especially dangerous and precariously at risk. Many governments, leaders, and ordinary people have already evolved beyond that level, others are trying to but struggling, still some see authoritarian government as the best form of government, because that is where they are in their development. Clear-thinking people at every level of culture and education all over the world struggle with the viciousness of authoritarian leaders, and view them with profound disappointment, feeling that they could have and should have done better, been more evolved, been better human beings. Sad to say, they're doing their best as they see it; in the stumbling progress of evolution, that is simply where they are.

The human urge *to be right* evolved before the capacity to determine truth did. It is common for people at all levels of education to want to be *right*. It is far less common for people to be totally committed to rigorously determining what is true. Admittedly, that determination can be complicated by brainwashing, lies, emotions, delusions, tricks our minds play, and seemingly solid, but actually false arguments. The desire to be right underlies belief's formidable offensive and defensive arsenals including tales of long-suffering efforts to find truth that lead predictably back to the belief.

Adding to the problem of the "need to be right" is the huge problem that all that we know is all that we know. We think things through and seek truth by surveying all that we know and seeing what that tells us. We feel that we think clearly and carefully, that we know a lot, that we give proper credence to the family members, communities, and experts we accept, and so doing, that our conclusions will be right, what we believe will be the truth. The problem is that our lives tend to be enclosed in a bubble of people who think the same way, believe the same things, rely on the same experts, accept the same *proofs* even if they are flawed, and support each other in overcoming doubts. Also, the psychological pressure of being told we're right or wrong by everyone around us is very strong. If our approach is unable to free itself from that bubble of belief and its *apparent logic*, if our approach is insufficient to the task for whatever reasons, and there can be many, we will not find what is actually true.

A further complication is our dreams. Our dreaming and spending a third of our lives asleep probably mainly developed during the 150 million years that we were nocturnal, lacked color vision, and lived in burrows. Then, life was simpler and out in nature where we had a lot to worry about. In many of our dreams, it seems like the old preverbal part of our mind from that time is struggling with our current language, color vision, and modern lives, and still tends to worry a lot. Often our literal wording as we describe such dreams reveals their meaning, and when it does, we find it obvious—after all, dreams from an old less evolved layer of our mind still came from *us*. Some of our dreams seem to come from more modern parts of our brain, at current mind-level, in full color, and amazing detail that seem completely real. It is easy to see how those who do not understand evolution could think supernatural forces are at work.

Our most powerful tools for finding truth are intuitive thinking, which is pattern thinking, and logic, which properly connects idea to idea. Intuitive thinking allows us to predict possibilities and likelihoods of things that may be true, but does not necessarily cover all the possibilities or provide exact likelihoods. Logic can determine cause and effect and truth, but invariably begins with something presumed to be true. For truth to actually result, that starting point, the thing considered to be true needs to be eminently verifiably scientifically true. Also, even educated experienced adults often lack a good knowledge of the scientific evidence that exists and the processes for determining that a theory is true, and for refining it in greater detail over time.

As science advances, old beliefs are being explained away one by one. We learn that pork made people sick and die because it contained a parasite. We learn that nothing we did made the volcano erupt, and nothing we did made it stop. We learn that malaria is not caused by harmful air or odors. We learn that miracles don't hold up under observation by skilled magicians and never include anything as simple as restoring the missing last digit of a finger. Sadly, a consequence of the human need to be right is that solid evidence that a belief is false rallies believers to *evade* that conclusion, demand the truth of their beliefs, and get more believers to join up.

Incomprehensible Us

We exist in what can be called the advancing now. The past exists only as features of this now that it crafted, the future exists only in what we and the forces of nature are making it to be. The life in each of us is billions of years old and has never known death. If it had, we could not have been born. That is because the life in each of us is the life of the living cell from our mother from which we developed. The life in that cell was the life in a cell that came from her mother. The life looking out of our eyes and her eyes and her mother's eyes is the same life. Our life has looked out of the eyes of a chain of mothers that goes back to the time before eyes, sex, or cells evolved, to a time where the features of life arose from molecules. Our experiences are different from all those mothers, but our life is the same life. Our experience-self sees tens of years; our life-self sees billions!

Ultimately a property of evolving light-like stuff, our life is also the same life that is in all the animals and plants, the life that sleeps in the stars, planets, and moons awaiting the conditions it needs to awaken. There are countess living things, but there is only one life.

We are profoundly and astonishingly important to and interconnected with everyone and everything around us, and the entire history of the earth before we were born, as we live our lives, and after we die. The critical event is our conception. A single one of 300 million sperm finds the egg. That particular sperm, that particular egg, at that particular moment in time depended on the whole evolution and history of the world including practically everything in the lives of our parents up to that moment. Change any of countless things, the life of an ancestor, interactions with family or friends, world affairs, health, public reaction to a passing comet, and on and on and on, and that sperm finding that egg at that moment would not have happened. Change history in any of a vast number of great or small ways prior to our conception and our particular sperm particular egg particular moment would not have been. Our life, our precise mixture of DNA, would not have happened, we would never have been born.

An enormous number of things, small things, ordinary things, great things and terrible things in history, prehistory, and even cosmic history had to happen just the way they did from the beginning of our universe for each of us to be born. They did not happen that way so that we would be born, but we were born because they happened that way.

Some may see themselves as unimportant but that doesn't make it so. Each and *every* single one of us is tremendously important. Because of our vast amazing interconnectedness, how we live our lives, our example, our kindness and unkindness, like stones dropped in a pond, send out waves of far-reaching effect on the births, whole lives, and deaths of countless others and the whole future of mankind and the earth.

Regardless of how we live our lives, the world will go on. We can be good or bad, we can give or take, help or hurt countless lives in ways we will usually never know; we choose. All *we* have is the quality of our lives and the lives of those we love, which is separate from our wealth, fame, or comfort, but more about truth, love, kindness, compassion, empathy and humanity. That has been well understood even from ancient times. Enormous numbers of good people whose kindness, integrity, and strength have made the world better for all around them, whose lives have never been recorded or remembered tower above those whose ambition was or is to be the ones who tower above.

As we learn the scientific facts of our evolution, aware of our submicroscopic features and the dust and molecules of the air, and aware that we are continuously taking in and pouring out molecules and energy, we discover that we are incredible storms of light and atoms, of molecules and energy, born of the entire past, affecting the entire future of this one vast living aware advancing evolving now, our cosmos of raging storms, great and small, unimaginably huge and unimaginably miniscule, of light and atoms, molecules and energy, old and new.

Look Look Look

Evolution pervades. Those who stand and fight, those who flee, minds that see the forest, minds that see the trees, who organize and simplify, who delve into detail, all come from survival traits of societies. Those who followed the herds minimized and organized their belongings *to survive*; those on the seashore gathered and kept everything that washed up that might be useful *to survive*. When descendants of those two lifestyles marry, those traits can conflict. The descendant of the nomads wants the home clean, organized, simplified. The descendant of the seashore only knows how to strew, stack, fill, and go get more. They have no idea what is really going on, or how to solve it.

Because we were originally four legged, our lower back wasn't well-designed to handle the stress of our upper body leaning this way and that as we stand on two legs, hence our common lower back injuries. For the same reason our insides are attached to our backs, but not our fronts, so standing on two legs, especially as we age, they sag downward. Plainly, our feet have evolved away from grasping and toward faster running; some toes can still grasp, but others only wiggle. Milk allergies, grain allergies, sickle cell anemia, the tendency to walk miles around where we are living, the tendency to stick close to home, the joy of physicality, relaxation, singing, dancing, and listening to a story all go back to eons of lifestyles' and environments' requirements for survival.

We see *enormous* human, animal, ecological, planetary and cosmic evolution spanning space and time.

Evolution of Morality

To succeed at anything: Do the best thing for the best reason in the best way at the best time. That is also the challenge to our morality, to manage our evolved traits in ways that are most highly evolved, kind, considerate, helpful, loving, and productive, and that support our well being, that of others and the earth. Failure and immorality happen, as those various "bests" become "not best" and "worst." It is tragic when people behave in primitive ways that violate the rights and lives of others, but that is a reality of our evolving species that has not yet evolved enough to be past such problems.

Our Cosmos

Not the only, but one of the real motivations for religion and philosophy is the powerful sense, as we view the heavens and the beauties and wonders of our earth, that we are part of something unimaginably vast, wonderful, and incomprehensible. Our mind-expanding time-stopping reaction is not just us being emotional, it's a genuine resonance. That is because *we* can look at the cosmos and realize what we are seeing. In that act, we are the dawning self-awareness of our universe, this advancing evolving now, this one being, this one vast life finding itself. So when our life-self looks out at all of it and says, "So this is what I am!" we feel it! That awesome resonance is both our cosmic self-discovery and our cosmic self-recognition!

We also have strong feelings of resonance with environments that were the whole world to our life-self for tens or hundreds of millions of years: as we see or smell the sea, swim, walk in a forest or grassland, climb mountains, crawl into a cozy place like a burrow, or sit high up in a tree. It feels like, *seems* like our life-self is recognizing and remembering its old homes.

What can we know about where our cosmos came from and where it is going? We know that it began with light-like stuff. We know that this one living being includes an evolution from inanimate to animate, from primitive mind to advanced mind, from basic instincts and survival to higher thought, logic, creativity, art, science, empathy, and love. We know that all life is the same life, that we are immediate family with all life on earth, with the earth itself, and our entire universe.

Eternal Now

Many people seek hope in various mythologies promising life after death. But the truth is, our common sense expects that when we die we are simply gone, stopped short at the moment of death. The problem is that believability of our ancient often tribal myths decreases as we learn, see, and understand more and more, and at the same time, we can't imagine any alternative to black nothingness that seems to really make any sense either. However, our lack of knowledge and imagination has no bearing on what is really the case. Consider how many things we now know that we could never have imagined.

We could never have imagined our universe that overwhelms us with its beauty, scale, and diversity. Yet, here it is. We would more likely have imagined "black nothingness!"

We didn't imagine that our earth is a ball, or that stars are gigantic fireballs of nuclear fusion, or that our naked eyes never see stars as they look now, but rather, because of their distances from earth, we see light that left them between 4.4 and 16,348 years ago—except for that from our blurry sister galaxy Andromeda which left her 2.5 million years ago.

We did not imagine that the stars are so tremendously far away that, except for the slow rotation of the sky, we seem to be stationary in space, when in fact we are speeding around the sun at 67,000 miles per hour. And who would imagine that many shooting stars are tiny grains of ice that zoom through our night sky not only because of their extreme velocity but also because of ours?

We could never have imagined this evolution, or the spans of time and processes involved. But our intellect, our scientific progress has shown it to us. We would more likely imagine, as we actually did, invisible good and evil spirits, gods and devils to explain the acts of nature, disease, luck, and good and evil in the world, as well as our very existence. In the evolution of thoughts about gods, there are still large populations with many spirits and gods, few spirits, one god, nature gods, authoritarian gods, fickle gods, and perfect eternal infinite loving gods who will burn us in the fires of hell forever unless we believe in, worship, and obey them!

17

Most people still do not imagine that the laws of science, which are the laws of nature, are *never* broken. Rather, unaware of how powerfully a sequence of random rewards affects our brains, most believe in gods, and spirits who *sometimes* answer our prayers, and that our chances of being answered probably increase if we pray very humbly and intensely, kneel, make promises or sacrifices, enlist the aid of a holy person, repeat devout mystical mantras, or pray in groups.

We could never have imagined that there is only one life on earth, that our life-self has been sensing its environments in ever increasing ways for *billions* of years, or that we are our universe's dawning self-awareness. Yet, all of that is true. We might more likely have imagined that the life within each of us began with us and ends with us, that our significance is whatever the god who ignited our personal spark of life wants it to be, or perhaps that we simply don't matter at all.

We could never have imagined the truly wondrous evolution of our cells, of our RNA and DNA, or the mixing of our genes in sexual reproduction, and how the resulting broad overlapping arrays of traits, skills, tendencies, and appearances of our males and females have made the human race so highly survivable and successful. Though some societies have accepted as natural, important, and equal all of the variations of good people, others aren't quite there yet, and still others, unable to recognize nature's survival wisdom, and ignorant of biological science cannot see beyond the philosophical and religious explanations, teachings, and rules of their revered but scientifically ignorant ancestors.

We didn't imagine the existence of bacteria or viruses, or the things about our minds and bodies that physics, chemistry, biology, medicine, psychology, and psychiatry have discovered. How could we?

Who would dream that we actually emit neutrinos that pass through our earth like it isn't even there!

And death? The fact that our common sense can't imagine how we may continue on as us beyond death has nothing to do with whether there is something good, something bad, or nothing. The fact that thinking, feeling, loving, and wanting to continue on somehow are all part of this amazing beautiful unimaginable being of which we are is at least a little bit encouraging.

A Brain in a Bone

Each of us is a brain in a bone, sensing the world around us with impressive but limited senses that are at, in, and not a millimeter beyond, the surface of our body. We feel that we certainly see that forest *way over there*, but that is an illusion, all we see is a pattern of activated rods and cones, immediately *on* the surface of our retinas.

Always, we see only a tiny fraction of the light that is present. As we look at the forest on a clear sunny day, we don't see the torrent of light that is beyond our visible spectrum, we don't see the brilliant flood of light from the sun *to* the forest, we don't see the light bouncing off the forest in all directions, we *only* see the few rays going from all the surfaces in the forest directly to our eyes, and even then, only *at* our retina surfaces.

Our bodies brightly emit infrared light; we don't see it. Our atmosphere is a storm of molecules, we don't see it, though sometimes, it can knock us down.

We don't see chlorophyll catching sunlight, the one source of all our energy.

We don't see the teeming hosts of little creatures smaller than 0.1 millimeter. As swallows swoop by, we see few or none of the tiny flying lives they are catching.

The space between the stars is brightly filled with visible and invisible starlight, but we see it as pitch black because we only detect the few visible rays that come *directly* to our retinas from each star.

Full Reality

We and all living things can only be truly understood when viewed as evolved and evolving cellular beings, which are part of a larger group of evolved and evolving molecular beings, which are part of a larger group of evolved and evolving atom beings, which are part of the tremendously largest group of all, evolved and evolving subatomic light-like stuff *beings.*

If we imagine light beings living in their light universe, we don't imagine us or rocks or trees, yet *we are* constructions of subatomic balled-up storms of light, living our lives in a surging spinning spiraling worldwide universe-wide complex electromagnetic sea of light, sensing and aware of only a small fraction of our complete selves and our total actual surroundings. The truth is: *We are light beings, constructed of light, powered by light, and this is our light universe.*

During earth's first billion years, the chemical and physical properties of atoms acted like a DNA for making molecules with the qualities of life. But a world of molecules of life is still a world of electromagnetism, light, energy, atoms, and inanimate molecules.

During earth's 2nd billion years, our ancestors, were prokaryotes (cells without nuclei) reproducing by simple cell division, some of whom evolved into eukaryotes, (cells with nuclei, like our cells) whose RNA and DNA continued evolving as life was finding its way to sexual reproduction, with its great survival benefits of gene mixing. Only the first cell that we are gets its DNA from our father, after that we develop and do repairs by old-fashioned cell division.

Earth-life

A thinking clam could look at the sea and the floor of the sea and know some things about the world it cannot imagine beyond the sea. Can we look at features of our world, space, time or ourselves and find aspects that extend beyond our familiar reality?

If we really look at all of evolution, we see one vast being with one vast life emerging from a seemingly inanimate light-like origin, fighting its way forward from outer space to the sea to the land, over billions of years adapting, competing, and evolving to survive, *and* that we are the part of it that is most highly evolved, we who now add technology to evolution's tool box.

Directed by their DNA our trillions of cells operate all the inner workings of our bodies, living, reproducing and dying with no sense of us as the beings they comprise. We too live, reproduce, and die, but what are we comprising? What are we leading to?

Each evolutionary level was needed to construct a next higher level that could do more advanced things than could ever have been done before. For us, that is to think at far higher levels than any other animal, and more than that, to determine truth, to fashion tools, to build, to create, and to develop increasingly advanced technology. Since all life on earth is one life, the struggle of all of earth's living things to survive is the struggle of their common earth-life to survive. We have evolved to be the part of earth-life that is capable of protecting it from natural dangers, diseases, and perhaps soon, passing asteroids, as well as our own activities.

We may no longer need to evolve physically, as our technology overcomes environmental demands. We can swim underwater; we can fly; we can air-condition our homes. Still we must wonder, in coming eons, how will evolution use our technology?

It has taken our life-selves and earth-life 4.5 billion years to emerge and evolve to where we are now. Although our sun will burn for another 5 billion years, 1 billion years from now it will swell up and burn off our oceans. Shall we achieve interstellar travel in the next 1 billion years and get out of here? Shall we move to a moon of Jupiter to have 4 or 5 billion more years rather than 1? Shall we increase and decrease earth's orbit as the sun swells up and cools down?

What is required for earth-life to survive is us, our evolved level of thought, our technology. But the challenge is enormous. Astronomical distances are, really *astronomical!* They are so great that it is hard to imagine how we may ever actually succeed. With current technology, it would take 5 years to get to a moon of Jupiter, and 303,705 years to get to the nearest star. The required knowledge and technology are very far beyond where we are now. Still, that may be the fantastic future of our evolution.

Do amazing discoveries about space, time, reality, dark matter, our full nature as light beings, and how to travel among the stars lie ahead? What will they tell us about the origin of all that we see, evolution, life, death, and what lies beyond and before our universe? Like the thinking clam looking at the sea and the bottom of the sea, it is probable that everything we see extends beyond and before our universe. And though making any sense of that still overwhelms us, we can feel good about the fact that all that we see of our universe is terrific!

Our imagination doesn't like the idea of being dead and gone, and really can't come up with how we may continue on, or for that matter anything we'd actually like to do forever! However, since things we could never have imagined abound in our universe, one more thing that we can't imagine is perfectly OK; in fact, it's good; it fits right in. We may reasonably judge that there is ample room to hope for something great that our current knowledge and limited common sense can't even *imagine.*